SpringerBriefs in Fire

Series Editor

James A. Milke

For further volumes:
http://www.springer.com/series/10476

Code Consultants, Inc.

Fire Flow Water Consumption in Sprinklered and Unsprinklered Buildings

An Assessment of Community Impacts

 Springer

Code Consultants, Inc.
Saint Louis
MO
USA

ISSN 2193-6595   ISSN 2193-6609   (electronic)
ISBN 978-1-4614-8108-9   ISBN 978-1-4614-8109-6   (eBook)
DOI 10.1007/978-1-4614-8109-6
Springer New York Heidelberg Dordrecht London

Library of Congress Control Number: 2013943328

Printed on acid-free paper

Springer is part of Springer Science+Business Media (www.springer.com)

Foreword

Over the past 30 years, selected municipal water authorities have implemented strategies, including stand by fees and other policies, to recover costs for water consumed in fires in sprinklered buildings. Typically, these fees are not directly related to sprinkler fire flows but rather are recognition of the fact that these flows are not metered and thus, not accounted for in conventional water cost recovery mechanisms. In contrast, water consumption at fires at unsprinklered properties is typically not subject to fees nor metered at the hydrant. With the growing adoption of residential sprinkler ordinances in communities across the country, the National Fire Protection Association commissioned this study to assess the relative community impacts of water consumption in sprinklered and unsprinklered properties.

This study considered standard estimates of the amount of water expected to be used in various building types with and without automatic sprinkler protection during a fire condition and also estimated the water used per year for commissioning, inspection, testing, and maintenance of buildings with systems for each building type. The total amount of water anticipated to be used for fire protection was compared with fees in sample jurisdictions; methods were developed to calculate fire water fees that are proportional to the anticipated volume of fire water used.

The Foundation acknowledges the contributions of the following individuals and organizations to this project:

Technical Panel

Anthony Apfelbeck, City of Altamonte Springs, FL
Fred Brower, Insurance Services Office, Inc. (ISO)
Jeff Feid, State Farm Insurance
Dawn Flancher, American Water Works Association
Russell Fleming, National Fire Sprinkler Association
Marc Gryc, FM Global
Tonya Hoover, CAL FIRE
Ed Kriz, City of Roseville, CA

Gary Keith, NFPA staff liaison
Matt Klaus, NFPA staff liaison

Sponsor

National Fire Protection Association

Project Contractor

Will Smith, Erin Crowley, Code Consultants, Inc.

Preface

A study has been conducted to analyze the estimated total fire protection water used in various building types. The study considered the water used in buildings with and without automatic sprinkler protection during a fire condition and estimated the water used per year for commissioning, inspection, testing, and maintenance (CITM) of buildings with sprinkler systems. The anticipated water used for fire protection was compared with the fees in sample jurisdictions; methods were developed to calculate fire water fees that are proportional to the anticipated volume of fire water used.

The study provides a detailed analysis for calculating the fire water demand required in sprinklered and unsprinklered buildings. This report shows that in all scenarios studied, the calculated water used during a fire when a building has a sprinkler system is less than that of an unsprinklered building. Additionally, the analysis indicates that in most of the scenarios studied the fire water used during a fire in an unsprinklered building exceeds the total water used in an otherwise similar sprinklered building for both CITM and a fire condition. These findings conclude that the owner of an unsprinklered building receives the full benefit of unlimited water through the public water system in a fire scenario without an increased cost, while the owner of a sprinklered building pays for the water used for CITM and a means that will reduce the amount of water required from private water system during a fire condition. In both cases, the cost of the water is typically not differentiated between sprinklered and unsprinklered buildings regardless of the reduction.

Guidance on the volume of water is provided by the International Code Council (ICC) [1], various National Fire Protection Association (NFPA) codes and standards [2, 3], and Insurance Services Office (ISO) guidelines [4]. These documents define the required fire flow and duration based on the construction type, use or occupancy, and area of a building.

The anticipated fire water usage has been compared with the current fire water fees in six sample jurisdictions. The sample jurisdictions were selected based on populations, range of building types, and fee structures.

Surveys were conducted to determine the fee structure in each of the sample jurisdictions. Methods used to charge for fire water included: direct usage charge at a fixed rate based on metering, direct usage charge at a rate that varies by season based on metering, fixed tap fee, initial capacity charge based on water line size,

and monthly capacity charge based on water line size. Two of the sample jurisdictions included a discount for the installation of a sprinkler system. One of the
sample jurisdictions permitted sprinkler systems with less than 20 sprinklers to be
connected to domestic water lines without additional charges and the other allowed
a reduction in the capacity charge to one-fifth the normal capacity charge rate.

A characteristic set of buildings were developed to compare fire flow water
consumption and fire flow fees in each of the sample jurisdictions. The characteristic set of buildings included each of the following building types:

- Residential, One- and Two- Family Dwelling
- Residential, Up to and Including Four Stories in Height
- Business
- Assembly
- Institutional
- Mercantile
- Storage

The fire flows for each of the building types was calculated based on the IFC
[1], NFPA 1 [2], NFPA 13 [3], NFPA 13D [5], NFPA 13R [6], and ISO [4] guidelines. The IFC and NFPA 1 include fire flow requirements for both sprinklered
and unsprinklered buildings. The required fire flow for a building protected with
a sprinkler system is typically permitted to be reduced by 50 % for one- and two-
family dwellings and 75 % for buildings other than one- and two-family dwellings. Available studies of fire water usage in sprinklered and unsprinklered
residential buildings show the volume of water to be conservative and indicate a
reduction of water used in a sprinklered home to be approximately 90 % less than
that of an unsprinklered home.

The volume of water required for all of the various building types studied
ranged from 60,000 to 585,000 gallons for unsprinklered buildings and from
30,000 to 480,000 gallons for sprinklered buildings when calculated according to
the IFC and NFPA 1. With some exceptions, the ISO guidelines typically require
less water than NFPA 1 or the IFC for unsprinklered buildings. The ISO guidelines
indicate a minimum fire water volume of 500 gpm for 2 h or the volumes required
by NFPA 13, whichever is greater.

For light-hazard occupancies, such as residential, business, assembly and institutional, the volume of water required by NFPA 13, NFPA 13R, or NFPA 13D, as
applicable, was in the range of 182–7,500 gallons, which was significantly less
than the volume required by the IFC or NFPA 1 for a sprinklered building. In
some instances for occupancies with a greater hazard classification, such as mercantile and storage, the volume of fire water required by NFPA 13 was greater than
the volume required by IFC and NFPA 1.

The installation of automatic sprinkler systems has the potential to significantly reduce the amount of water needed during a fire condition. Automatic sprinkler systems, however, require water for commissioning, inspection, testing, and
maintenance (CITM) that would not be required for a building without an automatic sprinkler system. The water required for CITM was estimated for each of the

characteristic building types. Over a 100 year period the volume of water required per year for CITM ranged from a low of approximately 14 gallons for a one- and two-family dwelling to a high of more than 94,000 gallons for a covered mall.

The total anticipated fire water used per year was calculated for each of the characteristic building types based on the required volume of fire water, the probability of a fire, and, for sprinklered buildings, the average CITM. For instance, in a typical one- and two-family dwelling, the fire water used for sprinklered buildings is between 4 and 10 % of the fire water used for unsprinklered buildings. Sprinklered apartment buildings used approximately 30 % of the fire water used by unsprinklered apartment buildings. The water savings can be seen in several of the building types.

The fees passed to the end user can be categorized into two basic categories: those fees related to construction costs and fees related to the cost of the commodity. This report concentrated on the fees related to the cost of the commodity. However, both could be accounted for with slight modifications. The fees related to construction costs would be inclusive of but not limited to "tapping fees" and installation fees. Likewise, the fees associated with the commodity would be billed monthly, or billed as a one-time commodity charge, or "capacity charge."

To adequately relate a fee for sprinklered and unsprinklered buildings, the current fee structure was calculated for sprinklered buildings and redistributed to sprinklered and unsprinklered buildings based on the estimated quantity of fire water used for each.

The amount of water used in sprinklered and unsprinklered buildings was identified by:

- The volume of water required by IFC, NFPA 1, ISO, NFPA 13, NFPA 13R, and NFPA 13D for sprinklered and unsprinklered buildings.
- The quantity of water based on the probability of a fire by the total number of buildings in each of the occupancies.
- The estimated volume of water used in sprinklered buildings during CITM.

Fees relating to construction cost were not included since they have no direct relationship with the quantity of water to be used by the connection. Further, the fees associated with standpipes in non-sprinklered buildings were also not considered since the intent of this study was to compare the fire protection water used in sprinklered with non-sprinklered buildings. As such, non-sprinklered buildings have been assumed to be completely unprotected. The total fee structure for water in each of the jurisdictions has been assessed per the number of sprinklered buildings in each of the occupancies.

The compiled graph in Fig. 1 indicates that unsprinklered buildings use more water than sprinklered buildings when comparing a single building.

The volume of water required in one- and two-family dwellings is negligible when compared to other buildings within a community.

The Water Research Foundation documents that the daily indoor per capita water use is approximately 69.5 gallons [7]. The average CITM for a sprinklered house per year varies from 14 to 28 gallons per year. The analogy illustrates how little water is used to maintain a residential sprinkler system.

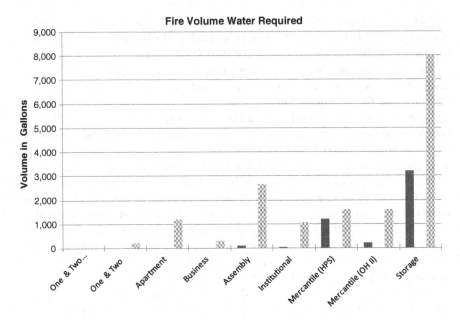

Fig. 1 Fire water required in gallons for sprinklered and unsprinklered buildings

■■■ = *Sprinklered*

▪▪▪ = *Unsprinklered*

In all cases, the volume of water required for a sprinklered building versus an unsprinklered building when comparing a single building is calculated to be less, based on the assumptions made. In most cases the total water required for CITM in a community with sprinklered buildings will be less than that required during a fire scenario for unsprinklered buildings.

Factors specific to a building type or low water pressures in a community may also impact fire water usage. When pressures supplied by a purveyor are low, the amount of water required by a fire protection system could be increased by the following factors:

- Low pressures may require larger pipe sizes to reduce pressure losses, which increases the volume of the systems and therefore the amount of water used in commissioning, testing, and flushing.
- Increased number of fire pumps which increase the volume of water used in CITM.

Fire water usage may also be influenced by the following factors based on building type:

- Local codes that require additional safety factors beyond those prescribed in NFPA 13.
- Malls require multiple tenants and have many systems to account for CITM.

- Large warehouses have many sprinkler systems and private fire service mains that require additional CITM. Some warehouses are governed by insuring authorities and require multiple sources (i.e. tanks or reservoirs).
- High rise buildings where multiple standpipe systems are required in conjunction with fire pumps.
- High rises that have no sprinkler systems, but are fitted with automatic standpipe systems.

Fire protection systems can be very complex, as such, each building should be assessed appropriately. The commodity of water is being sold to the end user. By virtue of this study, the only time water is used, is during a fire or during CITM. Therefore, if both are accounted for by volume, and fire service departments will use the water regardless of whether a building is protected by a sprinkler system or not, then the water used should be distributed to both buildings and not borne by the owner that provides a means for reducing the water used.

As stated before, fire water fees should be associated with those buildings that use the water for fire protection (This would include all buildings). As shown in the report, it would be appropriate to distribute the fees between both sprinklered and unsprinklered buildings.

The fee structure drafted for each community could differ; however, the basic concept is to charge fees for fire water that are proportional to the anticipated fire water used based on building type and on the presence of a fire sprinkler system. This report provides an estimate of fire water used for both fire conditions, including CITM, to allow communities to develop fire water fees for both sprinklered and unsprinklered buildings that are proportional to the anticipated fire water usage.

References

1. International Code Council. International fire code. Country Club Hills: International Code Council, Inc; 2012.
2. National Fire Protection Association. NFPA 1, fire code. Quincy: National Fire Protection Association; 2012.
3. National Fire Protection Association. NFPA 13, standard for the installation of sprinkler systems. Quincy: National Fire Protection Association; 2010.
4. Insurance Services Office. Guide for determination of needed fire flow. Jersey City: ISO Properties, Inc; 2008.
5. National Fire Protection Association NFPA 13D, standard for the installation of sprinkler systems in one- and two-family dwellings and manufactured homes. Quincy: National Fire Protection Association; 2010.
6. National Fire Protection Association NFPA 13R, standard for the installation of sprinkler systems in residential occupancies up to and including four stories in height. Quincy: National Fire Protection Association; 2010.
7. Water Research Foundation. Residential end uses of water. Denver: AWWA Research Foundation and the American Water Works Association; 1999.

Contents

Chapter 1
Introduction

Water used by fire protection systems is typically provided by a local purveyor. The local purveyor distributes water for use in residential, commercial and industrial buildings through delivery systems that usually include elevated gravity tanks, ground storage tanks and pumping systems, or a combination of both. Water is provided to the end user for a fee charged by the local purveyor.

According to the United States Geological Survey (USGS), [8] an estimated 410 billion gallons of water per day (Bgal/d) where used in the United States based on data from 2005. Of that, the leading users were thermo-electric power generation and irrigation. Public use was estimated at 44.2 Bgal/d, approximately 13 % of all fresh water used in a day, and 21 % of all freshwater used excluding thermo-electric power generation. The amount of water used by the public increased from 2000 to 2005 by 2 %, while the population increased by more the 5 %.

Water used for fire protection is a fraction of the overall public use of water. While fire protection systems have an impact on water use, this study is provided to help water purveyors and others evaluate water usage for fire protection and provides perspective on the amount of water used for fire protection compared with other uses.

Fire water flow requirements for buildings in the United States are typically based on model codes and standards published by the National Fire Protection Association (NFPA) or International Code Council (ICC) as well as guidance from the Insurance Services Office (ISO). The information contained in this report is based on the required fire flow from the following documents:

- NFPA 1, Fire Code, 2009 Edition
- ICC, International Fire Code (IFC), 2012 Edition
- ISO, Guide for Determination of Needed Fire Flow, 2008 Edition
- NFPA 13, Standard for the Installation of Sprinkler Systems, 2010 Edition
- NFPA 13D, Standard for the Installation of Sprinkler Systems in One- and Two-Family Dwellings and Manufactured Homes, 2010 Edition
- NFPA 13R, Standard for the Installation of Sprinkler Systems in Residential Occupancies up to and including Four Stories in Height, 2010 Edition

Code Consultants, Inc., *Fire Flow Water Consumption in Sprinklered and Unsprinklered Buildings*, SpringerBriefs in Fire, DOI: 10.1007/978-1-4614-8109-6_1, © Fire Protection Research Foundation 2012

- NFPA 25, Standard for the Inspection, Testing, and Maintenance of Water-Based Fire Protection Systems, 2011 Edition

Actual water used for fire protection will differ due to many variables including: pressure, system design, fire department use and response time.

1.1 Background

Over the past thirty years, selected municipal water authorities have implemented strategies, including stand-by fees and other policies, to recover costs for water consumed in fires in sprinklered buildings. It is also used to fund maintenance of the distribution system, tanks, pumps and pipes necessary to get these higher than normal demands to the needed location. Typically these fees are not directly related to sprinkler flows but rather are in recognition of the fact that these flows may not be metered and thus not accounted for in conventional water cost recovery mechanisms. In contrast, water consumed at fires at unsprinklered properties is typically not subject to fees nor metered at the hydrant. With the growing adoption of residential sprinkler ordinances in communities across the country, it is appropriate to assess the relative community impacts of water consumption in sprinklered and unsprinklered properties.

1.2 Research Objective

To assess the current prevalence and structure of fire flow fees against the community impact/water usage at sprinklered and unsprinklered properties to develop a consumption-based rational for community fire fighting resources.

1.3 Tasks

The following set of tasks is provided as a guide to help evaluate the use of water in fire protection systems.

1. Selection of at least six case study communities that traditionally have had fire flow fees, with a predetermined mix of building occupancies including residential, commercial and industrial.
2. Assessment of the fire flow fee structure in these communities as well as other nationally available information.
3. Select a characteristic set of both sprinklered and unsprinklered buildings within each community deemed to be representative of the building stock including residential, commercial and industrial.
4. Calculation of fire flow for sprinkler systems in these buildings as well as additional water consumption associated with sprinkler maintenance and testing.

5. Review previous documented literature to help make an assessment of water consumption for unsprinklered buildings in the case study communities.
6. Provide a rational basis for the assessment of fire flow fees for sprinklered and unsprinklered building in the selected communities.

Chapter 2
Community Selection

2.1 Population

Table 2.1 is a list of candidate communities categorized by population. The population indicated in the table below is based on the most recent data from the 2010 Census of Population and Housing produced by the U.S. Census Bureau.

The communities included in Table 2.1 represent a range of locations through the United States, including regions with very limited water availability and those with relatively plentiful water resources.

2.2 Building Types

The range of building types in each candidate community is summarized in Table 2.2, below, based on data from the U.S. Census Bureau. Table 2.2 does not include buildings in the following categories: information, real estate rental and leasing, administrative and support, and waste management and remediation services. The data included in Table 2.2 indicates that communities in each of the population categories are available that include each of the building types.

2.3 Water Fee Structure

In addition to the information on population and building types summarized in Tables 2.1 and 2.2, above, the water fee structure in each of the candidate communities was also considered. Water purveyors in each of the candidate communities were researched online and contacted directly to verify that the data collected was accurate. For many of the cities, especially the larger ones, water is provided

Code Consultants, Inc., *Fire Flow Water Consumption in Sprinklered*
and Unsprinklered Buildings, SpringerBriefs in Fire, DOI: 10.1007/978-1-4614-8109-6_2,
© Fire Protection Research Foundation 2012

Table 2.1 Candidate community populations [9]

Category	Community	Population
1. Population less than 20,000	Johnstown, OH	4,427
	Willmar, MN	17,926
	Rawlins, WY	8,633
	Rolla, MO	19,599
2. Population greater than 20,000 less than 100,000	St. Charles, MO	63,695
	Rockville, MD	59,825
	Bangor, ME	31,373
	Palm springs, CA	47,185
3. Population greater than 100,000 less than 500,000	Rochester, NY	208,001
	Roseville, CA	109,497
	Orlando, FL	227,961
4. Population greater than 500,000 less than 1,000,000	Denver, CO	600,158
	Tucson, AZ	520,116
	Ft Worth, TX	741,206
5. Population greater than 1,000,000	Los angeles, CA	3,792,621
	Philadelphia, PA	1,526,006

through multiple water departments; in those cases, only one water department for each community was researched for this report.

The six example communities were selected to represent a variety of water fee structures. The following features were represented in the water fee structures of the candidate communities:

- A fee structure designed to accommodate drought conditions;
- A simple fee structure based on a fixed cost per water connection; and
- A fee structure based on metering of water usage for fire protection systems.

Additional information on connection and tapping fees was reviewed for each candidate community. The concept of providing a credit to buildings protected with fire sprinkler systems was considered based on the premise that buildings protected with sprinkler systems would use less water in the event of a fire condition than fire department suppression efforts in a building not protected with sprinkler systems. However, this concept does not appear to be considered in the fee structures of most water purveyors.

All of the communities interviewed charge a fee for water service to fire sprinkler systems in commercial buildings. However, some communities do not charge an additional fee for water service to fire sprinkler systems in one- and two-family dwellings. Fire sprinkler systems protecting one- and two-family dwelling usually require only a small water service. This allows water service for fire sprinkler systems in one- and two-family dwellings to be provided by the normal domestic water line in some communities, which then charge for water usage at the normal domestic water rates without additional fees for fire sprinkler systems. Data from the selected communities is provided in appendix A.

Table 2.2 Candidate community building types (number of buildings per use) [10]

Community	Housing units	Manufacturing	Retail trade	Professional scientific and technical services	Educational services	Health care and social assistance	Art, entertainment, and recreation	Accommodations and food services	Other services (except public administration)
Category 1									
Johnstown, OH	1,928	18	32	19	3	11	5	13	23
Willmar, MN	8,127	32	146	64	1	139	9	63	52
Rawlings City, WY	3,861	0	53	23	0	34	2	42	28
Rolla, MO	8,110	21	150	42	5	121	7	90	45
Category 2									
St. Charles, MO	27,507	77	295	265	10	191	26	196	142
Rockville, MD	24,328	65	316	885	41	321	42	237	235
Bangor, ME	14,939	45	332	165	22	325	20	140	101
Palm Springs, CA	36,427	42	222	186	11	264	43	252	115
Category 3									
Rochester, NY	102,094	463	695	687	57	505	91	529	415
Roseville, CA	43,551	69	549	467	30	407	38	333	193
Orlando, FL	112,026	281	1,552	1,838	89	891	156	851	603
Category 4									
Denver, CO	274,061	840	2,271	3,876	256	1,981	290	1,778	1,627
Tucson, AZ	225,394	459	2,080	1,778	146	1,825	186	1,210	1,008
Ft Worth, TX	211,035	785	1,872	1,458	93	1,436	143	991	987
Category 5									
Los Angeles, CA	1,337,706	7,185	11,208	12,711	708	9,562	5,513	6,771	6,638
Philadelphia City, PA	661,299	946	4,420	2,756	254	3,600	350	3,396	2,466

2.4 Summary

Based on the methodology outlined above, the following six example communities were selected for further analysis:

- Johnstown, OH
- St. Charles, MO
- Orlando, FL
- Rochester, NY
- Denver, CO
- Los Angeles City, CA.

Data on population, location, building types, and water fee structures was considered as outlined below.

- Population and location
  - Candidate communities were divided into five categories based on population.
  - Example communities could not be located in a single area, but were selected to represent various regions of the United States.

- Building Types
  - Each community selected has a variety of residential, commercial, and industrial building types.
- Water Fee Structure
  Communities were selected with a variety of water fee structures, with examples that may address the following features:
  - Drought conditions
  - Fees per connection
  - Metered connections
  - Tapping fees.

Chapter 3
Assessment of Fire Flow Fees

All sixteen candidate communities were contacted to determine how fees are generated for the use of water in fire protection systems. This section provides information gathered from the six example communities selected for further analysis.

Representatives of the water purveyors for each of the candidate communities were interviewed by phone and asked the following set of five questions:

Question No. 1 Is there a meter charge for fire water service?
Question No. 2 Does the city have any drought conditions that would affect the cost of water used in fire protection systems?
Question No. 3 Do you have any incentives in place when a fire sprinkler system is provided?
Question No. 4 Is there an additional charge for water used in unprotected (by fire sprinkler systems) buildings?
Question No. 5 Is the there a fire district fee?

The following is the evaluation of each of the six water purveyors and how they collect fees for water used in fire protection systems:

3.1 Johnstown, OH [11]

Question No. 1 Is there a meter charge for fire water service?
Answer *Same as Domestic water rates see Table* 3.1
Question No. 2 Does the city have any drought conditions that would affect the cost of water used in fire protection systems?
Answer *No changes in fee due to drought conditions*
Question No. 3 Do you have any incentives in place when a fire sprinkler system is provided?
Answer *When a separate fire protection line is tapped the fee is 1/5 of the capacity charge*

Code Consultants, Inc., *Fire Flow Water Consumption in Sprinklered*
and Unsprinklered Buildings, SpringerBriefs in Fire, DOI: 10.1007/978-1-4614-8109-6_3,
© Fire Protection Research Foundation 2012

Table 3.1 Water rates

Fee	Gallons
$8.00	Minimum to 2,000 gallons
$3.60	3,000 gallons and beyond

Table 3.2 Tap and connection fees

Line size (inches)	Tap fee	Capacity charge*
¾	$375.00	$2,820.00
1	$375.00	$5,010.00
1 ½	$375.00	$11,280.00
2	$375.00	$20,050.00
3	$375.00	$45,110.00
4	$375.00	$80,200.00
6	$375.00	$180,450.00
8	$375.00	$320,800.00
10	$375.00	$501,240.00
12	$375.00	$721,790.00

*One-fifth the capacity charge will be charged if for fire protection only

Question No. 4 Is there an additional charge for water used in unprotected buildings?
Answer   *Hydrant water is not metered*
Question No. 5 Is the there a fire district fee?
Answer   *No additional Fee*

Table 3.1 reflects the water rates for metered water; however, they are the same rates as the domestic water. Table 3.2 indicates the tapping fees and capacity charge fees for water in fire protection systems, based on the size of the water service line.

Both the capacity charge and the tap fee are onetime fees and are independent of metered water rates. The water purveyor discounts the capacity charge when the water service is fire protection only, but still charges fees for water services for fire sprinkler systems that would not be charged for buildings without sprinkler systems.

3.2 St. Charles, MO [12]

Question No. 1 Is there a meter charge for fire water service?
Answer   *Fire service is not metered*
Question No. 2 Does the city have any drought conditions that would affect the cost of water used in fire protection systems?
Answer   *No changes in fee due to drought conditions*
Question No. 3 Do you have any incentives in place when a fire sprinkler system is provided?

Table 3.3 Water main connection fees

Water main connection (inches)	Fee
¾	$1,000
1	$1,600
1–1/2	$2,300
2	$3,900
3	$8,600
4	$15,400
6	$34,300
8	$41,400
10	$154,000
Unmetered main extension	$2,000
For sprinkler systems run off a domestic tap with more than 20 sprinklers	50 % increase

Answer   *No incentives for a sprinklered building*

Question No. 4   Is there an additional charge for water used in unprotected buildings?

Answer   *Hydrant water is not metered*

Question No. 5   Is the there a fire district fee?

Answer   *No additional Fee*

Table 3.3 outlines the fees charged for water main connections supplying fire protection systems.

A water main tap without a meter is $2,000 in addition to the tap fee based on the size of the connection. No additional fee is charged for a fire sprinkler system with less than twenty sprinklers that are tapped off of a domestic water line. There are no monthly fees for a fire water line, but buildings with a water line supplying a fire sprinkler system are charged initial connection fees that would not be charged to for buildings without sprinkler systems.

3.3 Orlando, FL [13]

Question No. 1   Is there a meter charge for fire water service?

Answer   *Yes, in addition to the monthly charge, if any water is used, it is charged at $1.54 per kilo gallon*

Question No. 2   Does the city have any drought conditions that would affect the cost of water used in fire protection systems?

Answer   *No changes in fee due to drought conditions*

Question No. 3   Do you have any incentives in place when a fire sprinkler system is provided?

Answer   *No incentives for a sprinklered building*

Question No. 4   Is there an additional charge for water used in unprotected buildings?

Table 3.4 Monthly charge (not including consumption charge)

Size of service (inches)	Inside city	Outside city
Less than 2	$9.70	$11.15
2	$9.70	$11.15
3	$9.70	$11.15
4	$9.87	$11.35
6	$20.96	$24.10
8	$40.08	$46.09
10	$68.84	$79.17
12	$108.71	$125.02
14	$227.09	$261.15
16	$405.16	$465.93

Answer   *Hydrant water is not metered*
Question No. 5   Is the there a fire district fee?
Answer   *No additional Fee*

A meter is provided on the incoming fire service line. A monthly fee shown in Table 3.4 is charged for the water service and an additional fee of $1.54 per thousand gallons is charged for any water used. No incentive is provided for the installation of a fire sprinkler system.

3.4 Rochester, NY [14]

Question No. 1   Is there a meter charge for fire water service?
Answer   *Yes, water is metered* and *billed monthly*
Question No. 2   Does the city have any drought conditions that would affect the cost of water used in fire protection systems?
Answer   *No changes in fire fees due to drought conditions*
Question No. 3   Do you have any incentives in place when a fire sprinkler system is provided?
Answer   *No incentives for a sprinklered building*
Question No. 4   Is there an additional charge for water used in unprotected buildings?
Answer   *Hydrant water is not metered*
Question No. 5   Is the there a fire district fee?
Answer   *No additional Fee*

Rochester, NY has two water mains available for use in fire protection. The first main provides fire protection through the domestic fire service. The second main is called a "Holly-High Pressure System". The fees billed depend on which system is used. Charges for the domestic fire service system and the "Holly High-Pressure System" are outlined in Table 3.5. Metered water consumption is also billed monthly and is summarized in Table 3.6.

Table 3.5 Monthly flat rates

Size of first check valve (inches)	Domestic fire service charge		Holly high-pressure system	
	Charge per quarter	Charge per month	Charge per quarter	Charge per month
2	$35.30	$11.77	–	–
4	$70.61	$23.54	$124.94	$41.65
6	$138.93	$46.31	$166.52	$55.51
8	$277.80	$92.60	$333.11	$111.04
10	$410.12	$136.71	$491.30	$163.77
12	$590.28	$196.76	–	–

Table 3.6 Metered water charge

Domestic fire service charge		Holly high-pressure system	
Gallons used per month	Charge per 1,000 gallons	Gallons used per month	Charge per 1,000 gallons
0–20,000	$3.01	0–20,000	$6.02
20,000–620,000	$2.77	20,000–620,000	$5.54
620,000–10,000,000	$2.17	Over 620,000	$4.34
10,000,000–15,000,000	$1.42	–	–
Over 15,000,000	$1.21	–	–

3.5 Denver, CO [15]

Question No. 1   Is there a meter charge for fire water service?
Answer   *Fire service is not metered*
Question No. 2   Does the city have any drought conditions that would affect the cost of water used in fire protection systems?
Answer   *No changes in fire fees due to drought conditions*
Question No. 3   Do you have any incentives in place when a fire sprinkler system is provided?
Answer   *No incentives for a sprinklered building*
Question No. 4   Is there an additional charge for water used in unprotected buildings?
Answer   *Hydrant water is not metered*
Question No. 5   Is the there a fire district fee?
Answer   *No additional Fee*

The monthly charge for a fire protection system water line is based only on the tap size. Water is not metered if used for fire protection and no additional fees are charge for water use through the fire water line. The fees for the monthly charge are shown in Table 3.7.

Table 3.7 Monthly charge

Tap size (inches)	Monthly charge
1	$4.68
2	$7.81
4	$12.08
6	$17.25
8	$30.19
10	$43.13
12	$60.00
16	$172.50
Hydrant	$17.25

3.6 Los Angeles City, CA [16]

Question No. 1   Is there a meter charge for fire water service?
Answer   *Yes, water is metered* and *billed monthly*
Question No. 2   Does the city have any drought conditions that would affect the cost of water used in fire protection systems?
Answer   *Los Angeles, CA, the city does not have a drought rate versus a normal rate. However, there is a fee billed for High Season and Low Season water use*
Question No. 3   Do you have any incentives in place when a fire sprinkler system is provided?
Answer   *No incentives for a sprinklered building*
Question No. 4   Is there an additional charge for water used in unprotected buildings?
Answer   *Hydrant water is not metered*
Question No. 5   Is the there a fire district fee?
Answer   *No additional Fee*

Table 3.8 Fire service monthly charge

Fire service monthly charge	
Size of fire service (inches)	Size of fire service (inches)
1 and smaller	$3.10
1−1/2	$11.00
2	$15.63
3	$38.49
4	$61.35
6	$108.48
8	$212.39
10	$255.79
12	$328.87
14	$511.58
16	$612.07
18	$821.03

Table 3.9   First tier rates (per hundred cubic feet of metered water)

High season		Low season					High season
Q1	Q2			Q3	Q4		
Jul–Sep	Oct	Nov	Dec	Jan–Mar	Apr	May	Jun
Meters under 2 inches							
$3.871	$3.840	$3.840	$3.840	$3.831	$3.681	$3.681	$3.681
Meters 2 inches and larger							
$3.871	$3.840	$3.840	$3.840	$3.831	$3.681	$3.681	$3.681

Table 3.10   Second tier rates (per hundred cubic feet of metered water)

High season		Low season					High season
Q1	Q2			Q3	Q4		
Jul–Sep	Oct	Nov	Dec	Jan–Mar	Apr	May	Jun
$5.69	$5.83	$5.83	$5.83	$5.84	$5.91	$5.91	$5.91

Buildings in Los Angeles are charged a monthly fee based on the size of the incoming fire service line, as outlined in Table 3.8, in addition to a metered water rate. The metered water rate is billed depending on the location of the building (low, medium, or high Temperature Zone) and the month that the water is used (low season versus high season). Additionally, water is charged at a "First Tier" rate, Table 3.9, and a "Second Tier" rate, Table 3.10. Normal fees are billed on the First Tier Rate up to a set volume of water. The threshold is dependent upon the area and the location (low, medium, or high Temperature Zone) of the building. Once the maximum is exceeded the fees are billed at the Second Tier Rate.

Chapter 4
Characteristic Set of Buildings

A characteristic set of buildings was developed to compare the fire water consumption for each building type. Fire water consumption was also compared between sprinklered and unsprinklered buildings. Features of the characteristic set of buildings were identified to allow the required fire flow to be calculated based on the requirements of the codes, standards, and guide listed below:

- NFPA 1, *Fire Code*, 2009 Edition
- ICC, *International Building Code* (IBC), 2012 Edition
- ICC, *International Fire Code* (IFC), 2012 Edition
- ISO, *Guide for Determination of Needed Fire Flow*, 2008 Edition
- NFPA 13, *Standard for the Installation of Sprinkler Systems*, 2010 Edition
- NFPA 13D, *Standard for the Installation of Sprinkler Systems in One- and Two-Family Dwellings and Manufactured Homes*, 2010 Edition
- NFPA 13R, *Standard for the Installation of Sprinkler Systems in Residential Occupancies up to and including Four Stories in Height*, 2010 Edition
- NFPA 25, *Standard for the Inspection, Testing, and Maintenance of Water-Based Fire Protection Systems*, 2011 Edition

The characteristic set of buildings included each of the following building types:

- Residential, One- and Two- Family Dwelling
- Residential, Up to and Including Four Stories in Height
- Business
- Assembly
- Institutional
- Mercantile
- Storage

Each of the occupancies listed will vary in footprint and fire flow requirements, with or without fire sprinklers. It is evident that the there are hundreds of combinations for building occupancies. It would be difficult to identify all sets of buildings in every jurisdiction.

Code Consultants, Inc., *Fire Flow Water Consumption in Sprinklered and Unsprinklered Buildings*, SpringerBriefs in Fire, DOI: 10.1007/978-1-4614-8109-6_4, © Fire Protection Research Foundation 2012

Having identified a set of building occupancies, the model codes were researched to distinguish a building that can be provided with or without sprinklers. The building sets were associated by construction type, area limitation, occupancy and whether protected or unprotected.

It should be noted that not all buildings listed in this report may be designed without sprinkler protection. Buildings that require sprinklers are noted based on code requirements.

4.1 Residential Buildings, One- and Two-Family Dwellings and Manufactured Homes (NFPA 13D)

Residential buildings have become a hot topic since all model codes now require residential occupancies to be provided with fire protection. However, not all jurisdictions have adopted the code revisions. The code requirements and building size are listed in Table 4.1.

Since these buildings are not limited in size based on the fire protection requirements, we have selected two building footprints for this category. The significance of having two buildings of different size is to identify that there are different flow requirements for each building that will be identified later.

- One and Two Family Dwelling 1–2,000 sf
- One and Two Family Dwelling 2–5,000 sf

4.2 Residential Buildings, Residential Occupancies Up to and Including Four Stories in Height (NFPA 13R)

Apartment buildings are another form of residential buildings. However, these buildings are required to follow the same guidelines for inspection, testing and maintenance as commercial buildings. It will be important to understand the additional flow due to inspection, testing and maintenance requirements. For this study there will be three different building types identified in Tables 4.2, 4.3 and 4.4

Table 4.1 Code requirements—one and two family homes [17, 18]

One and two family homes → Construction classification: unprotected, combustible			
Description	IBC	NFPA	ISO
Occupancy	Residential, R-3	Residential, 1 and 2 Family	Residential
Construction type	Type VB	Type V (000)	Class 1, C-4
Protection	*Maximum Area (sf)*		
Sprinkler protected provided	Unlimited	Unlimited	N/A
No sprinkler protected provided	Unlimited	Unlimited	N/A

Table 4.2   Code requirements—apartment building 1 [17, 18]

Residential apartment buildings
→ Construction classification: unprotected, combustible

Description	IBC	NFPA	ISO
Occupancy	Residential, R-2	Residential	Residential
Construction type	Type VB	Type V (000)	Class 1, C-4
Protection	*Maximum Area (sf)*		
Sprinkler protected provided	21,000	21,000	N/A
No sprinkler protected provided	7,000	7,000	N/A

Table 4.3   Code requirements—apartment building 2 [17, 18]

Residential apartment buildings
→ Construction classification: 2-hour exterior bearing walls
• Unprotected, non-combustible ext. elements
• Unprotected, combustible interior elements

Description	IBC	NFPA	ISO
Occupancy	Residential, R-2	Residential	Residential
Construction type	Type IIIB	Type III (200)	Class 2, C-4
Protection	*Maximum area (sf)*		
Sprinkler protected provided	48,000	48,000	N/A
No sprinkler protected provided	16,000	16,000	N/A

Table 4.4   Code requirements—apartment building 3 [17, 18]

Residential apartment buildings
→ Construction classification: unprotected, noncombustible

Description	IBC	NFPA	ISO
Occupancy	Residential, R-2	Residential	Residential
Construction type	Type IIB	Type II (000)	Class 3, C-4
Protection	*Maximum Area (sf)*		
Sprinkler protected provided	48,000	48,000	N/A
No sprinkler protected provided	16,000	16,000	N/A

There were two different sizes of each apartment building that could be used for each of the different construction classifications. One size is for an unsprinklered building and the other is for a sprinklered building. To compare the amount of water used in a sprinklered vs. unsprinklered building, the building size will need to be consistent between the two; therefore, the buildings that are used for this occupancy will be as follows:

• Apartment Building 1–7,000 sf
• Apartment Building 2–16,000 sf
• Apartment Building 3–16,000 sf

Table 4.5 Code requirements—low rise, business building [17, 18]

| Business buildings | | | |
| → Low rise, construction classification: unprotected, noncombustible | | | |
Description	IBC	NFPA	ISO
Occupancy	Business, B	Business	N/A
Construction type	Type IIB	Type II (000)	Class 3, C-2
Protection	*Maximum Area (sf)*		
Sprinkler protected provided	69,000	69,000	N/A
No sprinkler protected provided	23,000	23,000	N/A

Table 4.6 Code requirements—high rise, business building [17, 18]

| Business buildings | | | |
| → High rise, construction classification: protected, noncombustible | | | |
Description	IBC	NFPA	ISO
Occupancy	Business, B	Business	N/A
Construction type	Type IA	Type I (332)	Class 6, C-2
Protection	*Maximum area (sf)*		
Sprinkler protected provided	Unlimited	Unlimited	N/A
No sprinkler protected provided	Not permitted	Not permitted	N/A

4.3 Business Buildings

For business buildings there are two (2) buildings evaluated. One is a low rise and the other is a high rise. The high rise building technically could not be built without sprinkler protection. However, it is important to show how much water is used in a high rise building with special fire protection requirements such as standpipes and fire pumps. Tables 4.5 and 4.6 identify the building type and maximum sf allowances for this building type.

For low rise business buildings there are two maximum size buildings allowed with or without sprinklers. Maintaining consistency between the two; only one size building will be used to fit both applications. For the high rise building application only one size building will be applied over a 10-story building. Each floor will be provided with automatic sprinkler systems and standpipes, designed in accordance with the given standards. The buildings are identified as follows:

- Business Building, Low Rise–23,000 sf
- Business Building, High Rise–10 Floors, 40,000 sf/floor

4.4 Assembly Buildings

As with the business buildings there are two (2) assembly buildings for much of the same reasons, one is a low rise and the other is a high rise. Assembly buildings are buildings used for the gathering of persons for purposes such as civic, social,

Table 4.7 Code requirements—low rise, assembly buildings [17, 18]

Assembly buildings
→ Low rise, construction classification: unprotected, noncombustible

Description	IBC	NFPA	ISO
Occupancy	Assembly, A-2	Assembly > 1,000	N/A
Construction type	Type IIB	Type II (000)	Class 3, C-2
Protection	*Maximum area (sf)*		
Sprinkler protected provided	28,500	25,500	N/A
No sprinkler protected provided	9,500	8,500	N/A

Table 4.8 Code requirements—high rise, assembly buildings [17, 18]

Assembly buildings
→ High rise, construction classification: protected, noncombustible

Description	IBC	NFPA	ISO
Occupancy	Assembly, A-2	Assembly > 1,000	N/A
Construction type	Type IA	Type I (332)	Class 6, C-2
Protection	*Maximum area (sf)*		
Sprinkler protected provided	Unlimited	Unlimited	N/A
No sprinkler protected provided	Not permitted	Not permitted	N/A

religious functions; recreation, food, or drink consumption or awaiting transportation [17]. Tables 4.7 and 4.8 identify the building type and maximum area allowances for this building type.

For low rise Assembly buildings, there are four (4) different building sizes identify. To maintain consistency with all building codes for sprinklered and unsprinklered buildings, the two buildings identified for a low rise and a high rise are:

- Assembly Building, Low Rise–8,500 sf
- Assembly Building, High Rise–10 Floors, 50,000 sf/floor

4.5 Institutional Buildings

Staying within the same parameters for Business and Assembly Occupancies, two (2) Institutional Buildings are identified. Tables 4.9 and 4.10 distinguish the building type and maximum sf allowances for this building type.

For low rise and high rise buildings identified above, the following building sizes were chosen for the report:

- Institutional Building, Low Rise–11,000 sf
- Institutional Building, High Rise–10 Floors, 60,000 sf/floor

Table 4.9 Code requirements—low rise, institutional buildings [17, 18]

Institutional buildings
→ Low rise, construction classification: unprotected, noncombustible

Description	IBC	NFPA	ISO
Occupancy	Hospital, A-2	Health Care	N/A
Construction type	Type IIB	Type II (000)	Class 3, C-2
Protection	*Maximum area (sf)*		
Sprinkler protected provided	33,000	33,000	N/A
No sprinkler protected provided	11,000	11,000	N/A

Table 4.10 Code requirements—high rise, institutional buildings [17, 18]

Institutional buildings
→ High rise, construction classification: protected, noncombustible

Description	IBC	NFPA	ISO
Occupancy	Hospital, A-2	Health Care	N/A
Construction type	Type IA	Type I (332)	Class 6, C-2
Protection	*Maximum area (sf)*		
Sprinkler protected provided	Unlimited	Unlimited	N/A
No sprinkler protected provided	Not permitted	Not permitted	N/A

Table 4.11 Code requirements—mercantile, mall building [17, 18]

Mall buildings
→ Construction classification: unprotected, noncombustible

Description	IBC	NFPA	ISO
Occupancy	Covered mall building	Covered mall building	N/A
Construction type	Type IIB	Type II (000)	Class 3, C-3
Protection	*Maximum area (sf)*		
Sprinkler protected provided	Unlimited	Unlimited	N/A
No sprinkler protected provided	Not permitted	Not permitted	N/A

4.6 Mercantile Buildings

There are many different types of mercantile buildings that vary from small boutiques, big box retailers and large malls. This report identifies two (2) building types yet will have three (3) mercantile buildings included. The three (3) buildings will have three very different fire protection systems designed for each. This will provide insight that not all buildings of the same size will have the same flow requirements.

The first building is a Mall; this building would not be allowed to be built without a sprinkler system. Table 4.11 provides information regarding the building type and construction for the covered mall.

The second and third mercantile buildings will look the same as far as the building type and classification. Table 4.12 will be the model used for the other two (2) mercantile facilities.

Table 4.12   Code requirements—mercantile buildings (not a mall) [17, 18]

Mercantile buildings
→ Construction classification: unprotected, noncombustible

Description	IBC	NFPA	ISO
Occupancy	Mercantile, M	Mercantile	N/A
Construction type	Type IIB	Type II (000)	Class 3, C-3
Protection	*Maximum area (sf)*		
Sprinkler protected provided	37,500	37,500	N/A
No sprinkler protected provided	12,500	12,500	N/A

Table 4.13   Code requirements—storage buildings [17, 18]

Storage buildings
→ Construction classification: unprotected, noncombustible

Description	IBC	NFPA	ISO
Occupancy	Storage, S-1	Storage, Ordinary Hazard	N/A
Construction type	Type IIB	Type II (000)	Class 3, C-4
Protection	*Maximum area (sf)*		
Sprinkler protected provided	52,500	52,500	N/A
No sprinkler protected provided	17,500	17,500	N/A

As stated previously, there will be three (3) mercantile buildings identified by this report as follows:

- Mercantile Mall Building–750,000 sf
- Mercantile Building 1, OH 2–12,500 sf
- Mercantile Building 2, HPS–12,500 sf

4.7   Storage Buildings

Storage facilities can vary widely in size, shape and commodities being stored. Two (2) facilities are modeled within this report. One that can be provided with and without sprinklers under specified conditions and another that will require sprinklers, such as a distribution type warehouse. Table 4.13 identifies the code requirements used for the smaller of the two facilities.

The second building is unlimited area storage and would be used to facilitate warehouse activities for distribution. This building would require the use of sprinkler systems to provide protection and is identified in Table 4.14.

The two (2) buildings used in this category are identified as follows based on the data provided above:

- Storage Buildings–17,500 sf
- Storage Building Warehouse–1,000,000 sf

Table 4.14 Code requirements—storage buildings warehouse [17, 18]

Storage building warehouse
→ Construction classification: unprotected, noncombustible

Description	IBC	NFPA	ISO
Occupancy	Storage, S-1	Storage, ordinary hazard	N/A
Construction type	Type IIB	Type II (000)	Class 3, C-4
Protection	*Maximum Area (sf)*		
Sprinkler protected provided	Unlimited	Unlimited	N/A
No sprinkler protected provided	Not permitted	Not permitted	N/A

Table 4.15 Characteristic set of buildings

Building type	Area (sf)	Sprinklers	Notes
One and two family dwelling 1	2,000	Yes	–
One and two family dwelling 1	2,000	No	–
One and two family dwelling 2	5,000	Yes	–
One and two family dwelling 2	5,000	No	–
Apartment buildings (1)	7,000	Yes	–
Apartment buildings (1)	7,000	No	–
Apartment buildings (2)	16,000	Yes	–
Apartment buildings (2)	16,000	No	–
Apartment buildings (3)	16,000	Yes	–
Apartment buildings (3)	16,000	No	–
Business building, low rise	23,000	Yes	–
Business building, low rise	23,000	No	–
Business building, high rise	40,000	Yes	10 floors, standpipes, pump
Business building, high rise	40,000	No	10 floors, standpipes, pump
Assembly building, low rise	8,500	Yes	–
Assembly building, low rise	8,500	No	–
Assembly building, high rise	50,000	Yes	10 floors, standpipes, pump
Assembly building, high rise	50,000	No	10 floors, standpipes, pump
Institutional building, low rise	11,000	Yes	–
Institutional building, low rise	11,000	No	–
Institutional building, high rise	60,000	Yes	10 floors, standpipes, pump
Institutional building, high rise	60,000	No	10 floors, standpipes, pump
Mercantile building mall	750,000	Yes	Hydrants, pump, tenants
Mercantile building mall	750,000	No	Hydrants, pump, tenants
Mercantile Building 1	12,500	Yes	OH II
Mercantile building 1	12,500	No	OH II
Mercantile building 2	12,500	Yes	HPS
Mercantile building 2	12,500	No	HPS
Storage buildings	17,500	Yes	–
Storage buildings	17,500	No	–
Storage building warehouse	1,000,000	Yes	ESFR, Hydrants, 2 Pumps
Storage building warehouse	1,000,000	No	ESFR, Hydrants, 2 Pumps

4.8 Summary

Many building types and areas were defined in this section to help identify the Needed Fire Flow (NFF) requirements from the model building codes. This information is critical in providing continuity to both sprinklered and unsprinklered buildings for each. Additional buildings were added to this section that will demonstrate how much water is used in a sprinklered building with special fire protection requirements such as standpipes and fire pumps. It should be noted that this set of buildings, only provides information for use in this report, and should not be used to define the building characteristics in any jurisdiction. Table 4.15 provides a summary of all the characteristic set of buildings to be used for identifying flow requirements for sprinklered vs. unsprinklered buildings in this project.

Chapter 5
Calculation of Fire Water Volume

Calculation of required fire flows and volume for the model building codes for sprinklered and unsprinklered buildings is derived from IFC, Appendix B and NFPA 1, Sect. 18.4. Calculated fire flows for unsprinklered and sprinklered buildings are also derived from NFF through ISO and design densities or minimum sprinkler flows and volume required for sprinkler systems designed per NFPA 13, NFPA 13R and NFPA 13D. The volume of fire protection water is provided through a series of calculated flows (gpm) for a minimum duration (min) with detail shown in Appendix B.

Each of the Building Codes, Standards and ISO requirements were used to determine the correct flow, and take full advantage of any reductions of water allowed for the design of the systems.

The calculated fire flows for each of the building types is listed in Table 5.1, and reflects only the water required for a fire scenario in total gallons. All commissioning, inspecting, testing and maintenance, requirements are not provided in these calculations and shall be addresses in the next section. Detailed calculations used for the derivation of the calculated flows are provided for reference in Appendix B.

Code Consultants, Inc., *Fire Flow Water Consumption in Sprinklered*
and Unsprinklered Buildings, SpringerBriefs in Fire, DOI: 10.1007/978-1-4614-8109-6_5,
© Fire Protection Research Foundation 2012

Table 5.1 Calculated Fire Water Volumes [1, 2]

| Building type | Area (sf) | Sprks | Fire water volume (Gallons) | | | |
			IFC	NFPA 1	ISO	NFPA 13
One and two family homes 1	2,000	Yes	30,000	30,000	60,000	182
One and two family homes 1	2,000	No	60,000	60,000	120,000	N/A
One and two family homes 2	5,000	Yes	120,000	30,000	60,000	260
One and two family homes 2	5,000	No	240,000	60,000	180,000	N/A
Apartment buildings (1)	7,000	Yes	180,000	72,000	120,000	1,560
Apartment buildings (1)	7,000	No	270,000	270,000	240,000	N/A
Apartment buildings (2)	16,000	Yes	180,000	82,560	120,000	1,560
Apartment buildings (2)	16,000	No	330,000	330,000	240,000	N/A
Apartment buildings (3)	16,000	Yes	180,000	82,560	120,000	1,560
Apartment buildings (3)	16,000	No	330,000	330,000	180,000	N/A
Business building, low rise	23,000	Yes	270,000	146,340	60,000	5,700
Business building, low rise	23,000	No	585,000	585,000	240,000	N/A
Business building, high rise	40,000	Yes	270,000	168,840	60,000	5,700
Business building, high rise	40,000	No	N/A	N/A	450,000	N/A
Assembly building, low rise	8,500	Yes	180,000	72,000	60,000	7,500
Assembly building, low rise	8,500	No	240,000	240,000	180,000	N/A
Assembly building, high rise	50,000	Yes	360,000	255,120	60,000	7,500
Assembly building, high rise	50,000	No	N/A	N/A	450,000	N/A
Institutional building, low rise	11,000	Yes	180,000	72,000	60,000	5,835
Institutional building, low rise	11,000	No	270,000	270,000	180,000	N/A
Institutional building, high rise	60,000	Yes	360,000	270,000	60,000	5,835
Institutional building, high rise	60,000	No	N/A	N/A	540,000	N/A
Mercantile building mall	750,000	Yes	480,000	480,000	60,000	21,120
Mercantile building mall	750,000	No	N/A	N/A	1,440,000	N/A
Mercantile building 1	12,500	Yes	180,000	120,000	60,000	33,000
Mercantile building 1	12,500	No	270,000	270,000	180,000	N/A
Mercantile building 2	12,500	Yes	180,000	120,000	204,000[a]	204,000
Mercantile building 2	12,500	No	270,000	270,000	180,000	N/A

(continued)

Table 5.1 (continued)

Building type	Area (sf)	Sprks	Fire water volume (Gallons)			
			IFC	NFPA 1	ISO	NFPA 13
Storage buildings	17,500	Yes	180,000	120,000	132,000[a]	132,000
Storage buildings	17,500	No	330,000	330,000	300,000	N/A
Storage building warehouse	1,000,000	Yes	480,000	480,000	102,225[a]	102,225
Storage building warehouse	1,000,000	No	N/A	N/A	1,680,000	N/A

[a]ISO requirements are derived from NFPA 13 Design Criteria when the minimums are exceeded

Chapter 6
Calculation of Water Usage for Commissioning, Inspection, Testing, and Maintenance

Flow from fire protection systems for commissioning, inspection, testing and maintenance (CITM) are calculated for each building used in the study. The flows were derived for each phase and added to Table 6.1 for reference.

6.1 Commissioning

Flushing the mains if required per NFPA 13
Filling the system a minimum 3 times before turning the system over to the client

- 1st time—fill to check for leaks
- 2nd time—to fix any leaks
- 3rd time—to hydrostatically test the system.

Provide a single flow test per NFPA 13
Provide flow testing for a fire pump test if required
Flow standpipes if required.

6.2 Inspection

No Flows Required.

6.3 Testing

Flow switch test (twice a year)
Backflow prevention test
Pump Test

Code Consultants, Inc., *Fire Flow Water Consumption in Sprinklered and Unsprinklered Buildings*, SpringerBriefs in Fire, DOI: 10.1007/978-1-4614-8109-6_6, © Fire Protection Research Foundation 2012

Table 6.1 Calculated volume for CITM in sprinkler systems [19]

Building type	Area (sf)	Sprks	Year 1 (gallons)	Year 10 (gallons)	Average/ years 100 yrs (gallons)
One and two family homes 1	2,000	Yes	86	203	14
One and two family homes 2	5,000	Yes	176	293	15
Apartment buildings (1)	7,000	Yes	773	3,138	278
Apartment buildings (2)	16,000	Yes	1,003	3,498	299
Apartment buildings (3)	16,000	Yes	1,003	3,498	299
Business building, low rise	23,000	Yes	2,444	10,405	895
Business building, high rise	40,000	Yes	28,166	87,342	6,673
Assembly building, low rise	8,500	Yes	1,508	9,108	853
Assembly building, high rise	50,000	Yes	36,150	111,241	8,275
Institutional building, low rise	11,000	Yes	1,564	9,202	858
Institutional building, high rise	60,000	Yes	41,740	131,531	9,948
Mercantile building mall	750,000	Yes	453,415	1,272,765	94,618
Mercantile building 1 (OH II)	12,500	Yes	4,588	12,422	904
Mercantile building 2 (HPS)	12,500	Yes	8,297	38,604	3,404
Storage buildings	17,500	Yes	6,492	19,699	1,505
Storage building warehouse	1,000,000	Yes	228,314	890,279	74,377

Hydrant Test
Standpipe Test (every 5 years).

6.4 Maintenance

Obstruction investigation (every 5 years)

• Drain and fill the system.

The calculated volumes in Table 6.1 indicate the volumes of water required for years 1, 5 and the average gallons used per year over a 100 year period. This assumes the system will not be remodeled or that any additional repairs to the system are required. Each time a system is drained, it is required to be filled again which will add to the total flows

It should be noted that flow switches are not required per NFPA 13D. Additionally, NFPA 13D systems are not bound by the requirements of NFPA 25. However, the volume of water for a single flow test once a year is included in the calculation. Appendix C contains the calculated volumes of water for Table 6.1.

Chapter 7
Fire Probability and Frequency

U.S. Census Bureau building data and NFPA fire data was used to determine fire probability and fire frequency per occupancy for each jurisdiction surveyed.

7.1 U.S. Census Bureau

The total number of buildings in the United States, per occupancy, was obtained from census data, which originates from tax return information supplied from the Internal Revenue Service. This census data was compiled and categorized into seven occupancies as summarized in Table 7.1.

The source data used to obtain the above information can be found in Appendix D of this analysis.

7.2 NFPA Structure Fires by Occupancy

The NFPA report, *Structure Fires by Occupancy*, was referenced to determine the total number of fires per occupancy [22]. This report provides an estimation of the average number of fires, per year, for incidents reported to local U.S. fire departments during 2004–2008. This data was compiled and categorized into seven occupancies as summarized in Table 7.2.

The source data used to obtain the above information can be found in Appendix D of this analysis.

Code Consultants, Inc., *Fire Flow Water Consumption in Sprinklered and Unsprinklered Buildings*, SpringerBriefs in Fire, DOI: 10.1007/978-1-4614-8109-6_7, © Fire Protection Research Foundation 2012

Table 7.1 Number of buildings in the United States [20, 21]

Occupancy	Number of buildings
Residential, 1 and 2 family homes	76,313,410
Residential, apartment homes	30,549,390
Business	13,649,410
Assembly	1,980,406
Institutional	1,863,430
Mercantile	2,183,678
Storage	1,300,715

Table 7.2 Average number of fires per year [22]

Occupancy	Number of fires
Residential, 1 and 2 family homes	264,530
Residential, apartment homes	109,360
Business	6,705
Assembly	21,870
Institutional	7,300
Mercantile	12,895
Storage	31,510

7.3 Determining Fire Probability and Fire Frequency per Occupancy

7.3.1 Fire Probability per Year

Fire Probability describes the relative possibility that a fire will occur in any given occupancy. Fire Probability per occupancy was determined by comparing the total building data obtained from the U.S. Census Bureau and NFPA fire data.

For each occupancy, the average number of fires per year has been divided by the total number of buildings to determine the probability a fire will occur in that occupancy. For example, there are approximately 6,705 reported fires in Business

Table 7.3 Fire probability per occupancy

Occupancy	Fire probability (%)
Residential, 1 and 2 family homes	0.35
Residential, apartment homes	0.36
Business	0.05
Assembly	1.10
Institutional	0.39
Mercantile	0.59
Storage	2.42

occupancies each year. Dividing that number by the total number of Business buildings (13,649,410) results in probability of 0.05 %. Therefore, a fire will occur in approximately 0.05 % of all Business buildings each year.

Table 7.3 below depicts this fire probability calculation for each occupancy surveyed.

7.3.2 Fire Frequency

Fire Frequency describes the rate of occurrence (per year) that a fire could be expected per building for each occupancy. Similar to Fire Probability, Fire Frequency per occupancy was determined by comparing the total building data obtained from the U.S. Census Bureau and the NFPA fire data.

For each occupancy, the total number of buildings has been divided by the average number of fires per year to determine the frequency a fire will occur in that occupancy. For example, there are approximately 13,649,410 Business buildings in the United States. Dividing that number by the average number of fires in Business occupancies each year results in a frequency of approximately 2,036 years. Therefore, every Business building will experience a fire approximately every 2,036 years.

Table 7.4 below depicts this fire frequency calculation for each occupancy surveyed.

7.4 Number of Fires per Occupancy in Each Jurisdiction

Fire Probability, as determined in Section C above, has been applied to each jurisdiction to approximate the number of buildings which will experience a fire each year. For example, the probability that a fire will occur in a Business building is 0.05 %. There are approximately 4,082 Business buildings in Orlando Florida, therefore it is expected that a fire will occur in 3 Business buildings per year in Orlando.

Table 7.4 Fire frequency per occupancy

Occupancy	Fire frequency (years)
Residential, 1 and 2 family homes	288
Residential, apartment homes	279
Business	2,036
Assembly	91
Institutional	255
Mercantile	169
Storage	41

Table 7.5   Number of buildings with fires per occupancy for each jurisdiction per year

Occupancy	Johnstown, OH	St. Charles, MO	Rochester, NY	Orlando, FL	Denver City, CO	Los Angeles City, CA
Residential, 1 and 2 family homes	5	61	172	163	528	2,196
Residential, apartment homes	2	29	187	231	431	2,653
Business	1	1	1	3	5	17
Assembly	1	3	8	13	26	144
Institutional	1	1	2	4	8	38
Mercantile	1	2	5	10	14	121
Storage	1	2	12	7	21	175

Table 7.5 below depicts the fire probability calculation, per occupancy, for each jurisdiction surveyed per year.

Chapter 8
Compiled Data (Probability and Volume)

All data collected for probability, the number of building in each jurisdiction by occupancy, as well as the data calculated for fire flows and CITM are compiled into tables and are provided in Appendix E for reference. Each table represents a comparison of data as follows:

IFC Unsprinklered versus IFC Sprinklered.
NFPA 1 Unsprinklered versus NFPA 1 Sprinklered.
ISO Unsprinklered versus NFPA 13, 13R, 13D Sprinklered.
IFC Unsprinklered versus NFPA 13, 13R, 13D Sprinklered.
NFPA 1 Unsprinklered versus NFPA 13, 13R, 13D Sprinklered.

Two additional tables were provided in the One- and Two-family Dwellings that include Actual data collected in the Utiskul and Wu report [23] and the Bucks County Report [24].

Utiskul and Wu report Unsprinklered versus NFPA 13D Sprinklered.
Utiskul and Wu report Unsprinklered versus Bucks County, PA Report Sprinklered.

The data was evaluated based on the flows required for each of the occupancies in each of the model building codes.

8.1 Required Fire Volume Unsprinklered Buildings

Almost all IFC and NFPA 1 occupancies require the same volume of water for unsprinklered buildings with the exception of the example for One- and Two-Family Dwellings 2. The ISO process requires less water than either of the codes for unsprinklered buildings. With the exception of the one example, it should be noted that both the IFC and NFPA 1 are interchangeable. Having stated that, the most conservative model building code, or the code that requires the most water for an unsprinklered building, will be used to represent those flows required for unsprinklered buildings.

Code Consultants, Inc., *Fire Flow Water Consumption in Sprinklered
and Unsprinklered Buildings*, SpringerBriefs in Fire, DOI: 10.1007/978-1-4614-8109-6_8,
© Fire Protection Research Foundation 2012

8.2 Required Fire Volume Sprinklered Buildings

NFPA 13, 13R and 13D provide additional requirements on the minimum flows required by sprinkler systems to contain a fire. In all cases, the water required by NFPA 13, 13R, and 13D is considerably less than that required by the model building codes for sprinklered buildings. Additionally, the NFPA report *U.S. Experience with Sprinklers* provides fire sprinkler statistics and indicates that typically only one or two sprinklers are required to control a fire [25]. The report details that 88 % of reported fires involve one or two sprinklers. However, for this report, the minimum volume of water to be used should be those volumes required by a recognized standard. With this information, it was decided that the data evaluated for sprinklered buildings will be projected through the use of the required volumes from NFPA 13, 13R and 13D based on calculations and design densities.

8.3 Volume of Water Required

The water required for sprinklered buildings added to the water required for inspection testing and maintenance is provided in Fig. 8.1 for each of the occupancies. Likewise, the water required for unsprinklered buildings is provided in Fig. 8.2. These two graphs detail the water required based on; probability, hazard, occupancy and construction type as discussed within this report and will not apply to all building throughout all jurisdictions.

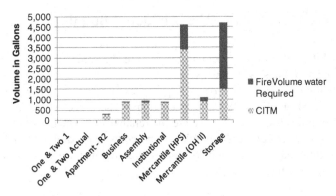

Fig. 8.1 Volume required sprinklered

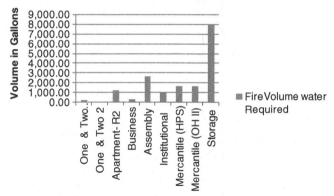

Fig. 8.2   Volume required unsprinklered

8.4 One- and Two-Family Dwellings

The data collected for One- and Two-Family Dwellings 1 is presented in Table 8.1 per the IFC and NFPA 13D.

The last column in each of the tables represents the percent of water used if all building were sprinklered over the water used if all buildings were unsprinklered. While theoretically this is not possible, the idea details comparing the total water used if all buildings were sprinklered as opposed to unsprinklered. In the case listed above, sprinklered buildings represent approximately 7 % of water used in unsprinklered buildings. For CITM the volume required assumes all buildings in each jurisdiction are sprinklered.

Table 8.2 represents the data calculated for actual fire data collected from the reports identified earlier.

It is interesting to identify that for Table 8.2 the percent of water used is at 124 %. This would indicate that at some point, sprinklered building versus unsprinklered buildings, the number of buildings sprinklered will use more water per year than unsprinklered building based on the CITM and the probability of a fire.

The data collected for One- and Two-Family Dwellings 2 is presented in Table 8.3 per the IFC and NFPA 13D.

In this table, the percent of water usage is down to 2 % because the amount of water used in unsprinklered buildings is double that of One- and Two-Family Dwellings 1.

8.5 Apartment Buildings

The data collected for Apartment Buildings 2 is presented in Table 8.4 per the IFC and NFPA 13R.

Table 8.1 One- and two-family dwelling 1 (IFC and NFPA 13D) [1, 5]

One- and two-family dwellings 1	Total buildings	Average number of fires per year	Gallons required unsprinklered building (IFC)	Gallons required sprinklered building (NFPA 13R)	Average gallons per year CITM for all buildings	Total gallons all buildings sprinklered	All sprinklered buildings/unsprinklered buildings (%)
Gallons/building	–	0.35 %	60,000	182	14	–	–
Johnstown, OH	1,389	5	288,887	876	19,446	20,322	7
St. Charles, MO	17,644	61	3,669,631	11,131	247,016	258,147	7
Rochester, NY	49,525	172	10,300,298	31,244	693,350	724,594	7
Orlando, FL	47,048	163	9,785,127	29,682	658,672	688,354	7
Denver City, CO	152,280	528	31,671,468	96,070	2,131,920	2,227,990	7
Los Angeles City, CA	633,460	2,196	131,748,148	399,636	8,868,440	9,268,076	7

Table 8.2 One- and two-family dwelling 1 (actual fire data)

One- and two-family dwellings 1	Total buildings	Average number of fires per year based on % buildings	Gallons required unsprinklered "residential fire sprinklers—water usage"	Gallons required "bucks county report"	Average gallons per year CITM for all buildings	Total gallons all buildings sprinklered	All sprinklered buildings/ unsprinklered buildings (%)
Flows/building	–	0.35 %	3,524	340	14	–	–
Johnstown, OH	1,389	5	16,967	1,637	19,446	21,083	124
St. Charles, MO	17,644	61	215,530	20,795	247,016	267,811	124
Rochester, NY	49,525	172	604,971	58,368	693,350	751,718	124
Orlando, FL	47,048	163	574,713	55,449	658,672	714,121	124
Denver City, CO	152,280	528	1,860,171	179,472	2,131,920	2,311,392	124
Los Angeles City, CA	633,460	2,196	7,738,008	746,573	8,868,440	9,615,013	124

Table 8.3 One- and two-family dwelling 2 (IFC and NFPA 13D) [1, 5]

One- and two-family dwellings 2	Total buildings	Average number of fires per year	Gallons required unsprinklered building (IFC)	Gallons required sprinklered building (NFPA 13)	Average gallons per year CITM for all buildings	Total gallons all buildings sprinklered	All sprinklered buildings/ unsprinklered buildings (%)
Gallons/building	–	0.35 %	240,000	260	14	–	–
Johnstown, OH	1,389	5	1,155,547	1,252	19,446	20,698	2
St. Charles, MO	17,644	61	14,678,523	15,902	247,016	262,918	2
Rochester, NY	49,525	172	41,201,194	44,635	693,350	737,985	2
Orlando, FL	47,048	163	39,140,510	42,402	658,672	701,074	2
Denver City, CO	152,280	528	126,685,871	137,243	2,131,920	2,269,163	2
Los Angeles City, CA	633,460	2,196	526,992,592	570,909	8,868,440	9,439,349	2

Table 8.4 Apartment buildings 2 (IFC and NFPA 13R) [1, 6]

Apartment buildings 2	Total buildings	Average number of fires per year	Gallons required unsprinkled building (IFC)	Gallons required sprinkled building (NFPA 13)	Average gallons per year CITM for all buildings	Total gallons all buildings sprinklered	All sprinklered buildings/ unsprinklered buildings (%)
Gallons/building	–	0.36 %	330,000	1,560	299	–	–
Johnstown, OH	449	2	530,416	2,507	134,251	136,758	26
St. Charles, MO	8,089	29	9,555,749	45,173	2,418,611	2,463,784	26
Rochester, NY	52,354	187	61,847,161	292,368	15,653,846	15,946,214	26
Orlando, FL	64,417	231	76,097,501	359,734	19,260,683	19,620,417	26
Denver City, CO	120,532	431	142,387,630	673,105	36,039,068	36,712,173	26
Los Angeles City, CA	741,201	2,653	875,600,287	4,139,201	221,619,099	225,758,300	26

8.6 Business Buildings

The data collected for Business Buildings is presented in Table 8.5 per the IFC and NFPA 13.

For business buildings the water required for fighting fires based on probability is significantly less than that for unsprinklered buildings. However, when the water required for CITM is added, the total water for sprinklered over unsprinklered buildings increases to 312 %. This would indicate that at some point, sprinklered building costs would be more than unsprinklered costs. However, the fact remains that the total fees will still be distributed to both buildings.

8.7 Assembly Buildings

The data collected for Assembly Buildings is presented in Table 8.6 per the IFC and NFPA 13

8.8 Institutional Buildings

The data collected for Institutional Buildings is presented in Table 8.7 per the IFC and NFPA 13D.

8.9 Mercantile Buildings

The data collected for Mercantile (HPS) Buildings is presented in Table 8.8 per the IFC and NFPA 13.

Again, based on the data represented above, the sprinklered building will eventually use more water per year based on CITM of the building than unsprinklered buildings by 289 %. Therefore, the fees for the sprinklered buildings will be more than the fees for the unsprinklered buildings at some point.

The data collected for Mercantile (OH II) Buildings is presented in Table 8.9 per the IFC and NFPA 13.

8.10 Storage Buildings

The data collected for Storage Buildings is presented in Table 8.10 per the IFC and NFPA 13.

Table 8.5 Business buildings (IFC and NFPA 13) [1, 3]

Business buildings	Total buildings	Average number of fires per year	Gallons required unsprinkled building (NFPA 1)	Gallons required sprinklered building (NFPA 13)	Average gallons per year CITM for all buildings	Total gallons all buildings sprinklered	All sprinklered buildings/ unsprinklered buildings (%)
Gallons/building	–	0.05 %	585,000	5,700	895	–	–
Johnstown, OH	118	0.06	33,910	330	105,610	105,940	312
St. Charles, MO	677	0.33	194,549	1,896	605,915	607,811	312
Rochester, NY	1,748	0.86	502,322	4,894	1,564,460	1,569,354	312
Orlando, FL	4,082	2.01	1,173,043	11,430	3,653,390	3,664,820	312
Denver City, CO	8,666	4.26	2,490,345	24,265	7,756,070	7,780,335	312
Los Angeles City, CA	33,340	16.38	9,580,901	93,352	29,839,300	29,932,652	312

Table 8.6 Assembly buildings (IFC and NFPA 13) [1, 3]

Assembly buildings	Total buildings	Average number of fires per year	Gallons required unsprinkled building (IFC)	Gallons required sprinklered building (NFPA 13)	Average gallons per year CITM for all buildings	Total gallons all buildings sprinklered	All sprinklered buildings/ unsprinklered buildings (%)
Gallons/building	–	1.10 %	240,000	7,500	853	–	–
Johnstown, OH	21	0	55,658	1,739	17,913	19,652	35
St. Charles, MO	232	3	614,885	19,215	197,896	217,111	35
Rochester, NY	677	7	1,794,298	56,072	577,481	633,553	35
Orlando, FL	1,096	12	2,904,801	90,775	934,888	1,025,663	35
Denver City, CO	2,324	26	6,159,450	192,483	1,982,372	2,174,855	35
Los Angeles City, CA	12,992	143	34,433,550	1,076,048	11,082,176	12,158,224	35

Table 8.7 Institutional buildings (IFC and NFPA 13D) [1, 5]

Institutional buildings	Total buildings	Average number of fires per year	Gallons required unsprinklered building (IFC)	Gallons required sprinklered building (NFPA 13)	Average gallons per year CITM for all buildings	Total gallons all buildings sprinklered	All sprinklered buildings/ unsprinklered buildings (%)
Gallons/building	–	0.39 %	270,000	5,835	858	–	–
Johnstown, OH	11	0	11,635	251	9,438	9,689	83
St. Charles, MO	191	1	202,026	4,366	163,878	168,244	83
Rochester, NY	508	2	537,325	11,612	435,864	447,476	83
Orlando, FL	891	3	942,435	20,367	764,478	784,845	83
Denver City, CO	1,981	8	2,095,357	45,283	1,699,698	1,744,981	83
Los Angeles City, CA	9,562	37	10,113,984	218,574	8,204,196	8,422,770	83

Table 8.8 Mercantile (HPS) buildings (IFC and NFPA 13) [1, 3]

Mercantile buildings (HPS)	Total buildings	Average number of fires per year	Gallons required unsprinklered building (IFC)	Gallons required sprinklered building (NFPA 13)	Average gallons per year CITM for all buildings	Total gallons all buildings sprinklered	All sprinklered buildings/ unsprinklered buildings (%)
Gallons/building	–	0.59 %	270,000	204,000	3404	–	–
Johnstown, OH	39	0.23	62,181	46,982	132,756	179,738	289
St. Charles, MO	295	1.74	470,347	355,373	1,004,180	1,359,553	289
Rochester, NY	695	4.10	1,108,106	837,236	2,365,780	3,203,016	289
Orlando, FL	1,552	9.16	2,474,504	1,869,626	5,283,008	7,152,634	289
Denver City, CO	2,271	13.41	3,620,876	2,735,773	7,730,484	10,466,257	289
Los Angeles City, CA	20,346	120.15	32,439,605	24,509,923	69,257,784	93,767,707	289

Table 8.9 Mercantile (OH II) buildings (IFC and NFPA 13) [1, 3]

Mercantile buildings (OH II)	Total buildings	Average number of fires per year	Gallons required unsprinklered building (IFC)	Gallons required sprinklered building (NFPA 13)	Average gallons per year CITM for all buildings	Total gallons all buildings sprinklered	All sprinklered buildings/ unsprinklered buildings (%)
Gallons/building	–	0.59 %	270,000	33,000	904	–	–
Johnstown, OH	39	0.23	62,181	7,600	35,256	42,856	69
St. Charles, MO	295	1.74	470,347	57,487	266,680	324,167	69
Rochester, NY	695	4.10	1,108,106	135,435	628,280	763,715	69
Orlando, FL	1,552	9.16	2,474,504	302,439	1,403,008	1,705,447	69
Denver City, CO	2,271	13.41	3,620,876	442,552	2,052,984	2,495,536	69
Los Angeles City, CA	20,346	120.15	32,439,605	3,964,841	18,392,784	22,357,625	69

Table 8.10 Storage buildings (IFC and NFPA 13) [1, 3]

Storage buildings	Total buildings	Average number of fires per year	Gallons required unsprinklered building (IFC)	Gallons required sprinklered building (NFPA 13)	Average gallons per year CITM for all buildings	Total gallons all buildings sprinklered	All sprinklered buildings/ unsprinklered buildings (%)
Gallons/building	–	2.42 %					–
Johnstown, OH	29	1	330,000	132,000	1505	–	59
St. Charles, MO	77	2	231,835	92,734	43,645	136,379	59
Rochester, NY	463	11	615,561	246,224	115,885	362,109	59
Orlando, FL	281	7	3,701,359	1,480,544	696,815	2,177,359	59
Denver City, CO	840	20	2,246,397	898,559	422,905	1,321,464	59
Los Angeles City, CA	7,185	174	6,715,208	2,686,083	1,264,200	3,950,283	59
			57,439,013	22,975,605	10,813,425	33,789,030	59

Table 8.11 Yearly average of water used in sprinklered and unsprinklered buildings

Building type	Sprinklered	Fire flow—NFPA 13, 13D, 13R over IFC (Gal/fire)	Fire probability (Fires/Yr) (%)	Yearly average fire flow (Gal/Yr)	Yearly average CITM (Gal/Yr)	Yearly average fire water usage (Gal/Yr)	Percentage of Sprinklered building over unsprinklered building (%)
One- and two-family homes 1	Yes	182	0.35	1	14	15	7
	No	60,000	0.35	210	0	210	
On- and two-family homes 2	Yes	260	0.35	1	15	16	2
	No	240,000	0.35	840	0	840	
Apartment buildings (1)	Yes	1,560	0.36	6	278	284	29
	No	270,000	0.36	972	0	972	
Apartment buildings (2)	Yes	1,560	0.36	6	299	305	26
	No	330,000	0.36	1,188	0	1,188	
Apartment buildings (3)	Yes	1,560	0.36	6	299	305	26
	No	330,000	0.36	1,188	0	1,188	
Business building	Yes	5,700	0.05	3	895	898	307
	No	585,000	0.05	293	0	293	
Assembly building	Yes	7,500	1.10	83	853	936	35
	No	240,000	1.10	2,640	0	2,640	
Institutional building	Yes	5,835	0.39	23	858	881	84
	No	270,000	0.39	1,053	0	1,053	
Mercantile building 1 (OHII)	Yes	21,120	0.59	125	904	1,029	65
	No	270,000	0.59	1,593	0	1,593	
Mercantile building 2 (HPS)	Yes	204,000	0.59	1,204	3,404	4,608	289
	No	270,000	0.59	1,593	0	1,593	
Storage buildings	Yes	132,000	2.42	3,194	1,505	4,699	59
	No	330,000	2.42	7,986	0	7,986	

8.11 Yearly Average

The yearly average for volume of water was calculated per building type, per building, based on the probability of a fire in the jurisdiction, for sprinklered and unsprinklered buildings. Then the percent of water was calculated per building type for sprinklered and unsprinklered buildings in Table 8.11 per the IFC and NFPA 13, 13D and 13R. It is interesting to note that the percent of water used in almost all cases are less than 75 %. Furthermore, the water used in one- and two-family dwelling is less than 10 %, while the water used in an apartment building might be less than 30 %.

The two anomalies over 100 % can be accounted for in the volume of water used in a fire and the number of buildings with recorded fires. For instance, the probability of a fire in a business building is extremely low; hence the amount of water calculated by probability for an unsprinklered building is low. However, the amount of water used per year in CITM is what drives the higher percentage. When reviewing the water used in a fire condition, the percentage is less than 2 %. For the other anomaly in a high piled storage configuration, the water used in a fire condition with a sprinkler system exceeds that of a fire condition without a sprinkler system. In this instance, IFC, NFPA 1 and ISO do not account for the hazard within the building as a part of the calculation.

Chapter 9
Calculation of Fire Fees

As discussed throughout the report, fees related to the water used in fire protection systems are typically based on the size of the fire service connection supplying the water to the building. The cost to the end user is billed either at one time "capacity charge" connection fee, a flat monthly fee, a metered water fee for a minimum volume of water or a combination of any of those listed. Therefore, the current total fees a purveyor charges can be assessed by the size of the mains installed to the site. The total fees currently collected by the purveyor will be designated as C_T. The fees that should not be included are; tapping fees, equipment fees or any related labor associated with the installation of the fire protection mains. Those fees are considered construction costs and are not related to the flows within the mains. C_T will be different for every jurisdiction. The cost of the current fee structure will provide a starting point to distribute the cost among sprinklered and unsprinklered buildings.

In theory, the total cost of the new fee structure should equal the cost of the existing fee structure per the equation below.

$$C_T = C_S + C_U$$

where

C_T Total cost of existing fee structure
C_S Cost of sprinklered buildings
C_U Cost of unsprinklered buildings

The above referenced fees are not directly proportionate. An added variable is provided to directly relate the fees to the quantity of water distributed between sprinklered and unsprinklered buildings.

The value of V_T or the total volume of water required for a sprinklered and unsprinklered building is provided. The volume of water was calculated previously in Section 6 for Sprinklered and Unsprinklered buildings and Section VII for CITM. Subsequently, the total volume is defined by the following equation.

Code Consultants, Inc., *Fire Flow Water Consumption in Sprinklered
and Unsprinklered Buildings*, SpringerBriefs in Fire, DOI: 10.1007/978-1-4614-8109-6_9,
© Fire Protection Research Foundation 2012

V_T Total volume of water required in gallons by probability

$$V_T = V_S + V_U$$

V_S Total volume of water required in sprinklered buildings in gallons by probability

V_U Total volume of water required in unsprinklered buildings in gallons by probability

Additionally, V_S and V_U can be further broken down into volumes of water in the equations below:

$$V_S = B_{W/S} * (F * V_{W/S} + V_{CITM})$$

where

$B_{W/S}$ Total number of sprinklered buildings in the occupancy

F Probability of a fire in the jurisdiction per the building type per year

$V_{W/S}$ Volume of water estimated in a sprinklered building during a fire in gallons

V_{CITM} Volume of water required CITM per year over a one hundred year period

$$V_U = B_{W/O} * F * V_{W/O}$$

where

$B_{W/O}$ Total number of unsprinklered buildings in the occupancy

$V_{W/O}$ Volume of water estimated in an unsprinklered building during a fire in gallons

Combining the two initial equations yields the final equation.

$$C_T * V_T = C_S * B_{W/S} * (P * V_{W/S} + V_{CITM}) + C_U * B_{W/O} * P * V_{W/O}$$

This is the equation that will be used to evaluate the fire fees in each of the jurisdictions. Slight modifications will be necessary to adapt the information to the style of billing for fire fees per the current fee structure in the jurisdiction. The equation above is slightly modified when the cost per connection of an unsprinklered building exceeds the original cost of a sprinklered building. To keep this from happening, a capped fee for the unsprinklered building was added at the original fee charged for a sprinklered building. The fees for the sprinklered building are then revised to account for the additional charges. This can be provided at the discretion of the water purveyor. If not provided, once all but a few buildings are not protected, the fees for the unsprinklered building become extremely high, in some cases, in the hundreds of thousands of dollars. Therefore, the graph shows a cap on the unsprinklered building to keep the costs manageable. As stated previously, this could be implemented at the water purveyor's discretion.

9.1 Johnstown, OH

Currently, Johnstown, OH charges a tapping fee, Capacity Charge, and a metered charge for the volume of water used. Tapping fees are a part of construction costs and therefore should be paid by the user. The Capacity Charge is a onetime fee and based on the size of the line entering the building. The Capacity Charge is directly related to the volume of water and should be distributed between both sprinklered and unsprinklered buildings as such:

$$C_T * V_T = C_S * B_{W/S} * (P * V_{W/S} + V_{CITM}) + C_U * B_{W/O} * P * V_{W/O}$$

The monthly fees would also be directly related to a volume of water and would be distributed to both sprinklered and unsprinklered buildings based on the number of buildings:

A set of graphs based on the number of buildings and fire fees for sprinklered and unsprinklered fees can be found in Appendix F. The volume of water used to develop the graphs was derived from IFC and NFPA 13, 13D, 13R and actual water data taken from the reports discussed.

9.2 St. Charles, MO

St Charles, MO charges a onetime fee for a water connection based on the size of the connection. The fee is based on the volume of water delivered to the site. There are no additional monthly fees for the use of the water. The fees for St Charles can be assessed as follows:

$$C_T = C_S * B_{W/S} * P * V_{W/S} + C_U * B_{W/O} * P * V_{W/O}$$

Since no water is charged after the initial connection for CITM, the flat fee is the total fee per building. A set of graphs based on the number of buildings and fire fees for sprinklered and unsprinklered fees can be found in Appendix G. The volume of water used to develop the graphs was derived from IFC and NFPA 13, 13D, 13R and actual water data taken from the reports discussed.

9.3 Rochester, NY

Rochester, NY has two separate systems that provide fire services. There is a Domestic Fire Service Charge and a Holly High-Pressure System. There is a monthly Service charge as well as a Consumption Rate. New meter fee and Tapping fee are also assessed; however, these fees are not a part of continuous fee for volume of water and are related to construction costs. The fee for CITM can be provided through the consumption rate for each system. The fees for Rochester, NY are assessed for both systems using the same equation below:

$$C_T * V_T = C_S * B_{W/S} * (P * V_{W/S} + V_{CITM}) + C_U * B_{W/O} * P * V_{W/O}$$

A set of graphs for both systems based on the number of buildings and fire fees for sprinklered and unsprinklered fees can be found in Appendix H. The volume of water used to develop the graphs was derived from IFC and NFPA 13, 13D, 13R and actual water data taken from the reports discussed.

9.4 Orlando, FL

Orlando, FL has two separate rates that provide fire services. There is an Inside City Rate and an Outside city rate. Both of the rates are assessed monthly. Additionally, there is a monthly Rate based on the kilo-gallons used during the month. There are no additional fees related to construction costs. The fees for Orlando, FL are assessed for both systems using the same equation below where the monthly fee can be assessed a minimum of twice a year for CITM:

$$C_T * V_T = C_S * B_{W/S} * (P * V_{W/S} + V_{CITM}) + C_U * B_{W/O} * P * V_{W/O}$$

A set of graphs for both systems based on the number of buildings and fire fees for sprinklered and unsprinklered fees can be found in Appendix I. The volume of water used to develop the graphs was derived from IFC and NFPA 13, 13D, 13R and actual water data taken from the reports discussed.

9.5 Denver, CO

Denver, CO has a fixed monthly fee based on fire service size only. There are no additional fees related to construction costs. The fees for Denver, CO are assessed using the equation below where the projected CITM is factored into the volume of water used per year and then distributed on a monthly basis:

$$C_T * V_T = C_S * B_{W/S} * (P * V_{W/S} + V_{CITM}) + C_U * B_{W/O} * P * V_{W/O}$$

A set of graphs for both systems based on the number of buildings and fire fees for sprinklered and unsprinklered fees can be found in Appendix J. The volume of water used to develop the graphs was derived from IFC and NFPA 13, 13D, 13R and actual water data taken from the reports discussed.

9.6 LA City, CA

LA City, CA has a fixed monthly fee based on fire service size. There is also a monthly service charge based on water used at a rate per HFC. The rates vary based on the size the service and the season in which the service is provided. Additionally, if more than a specified amount of water is used during the month, a second tier service amount is assessed. The equation below identifies the assessed fees using the first tier rate. If the maximum gallons of water are used by a fire protection system during CITM, then the fees for the second tier rate should only be applied to the owner of the system and not distributed to unsprinklered buildings. No additional construction fees are assessed. The volume of water for CITM should account for minimum flow each month as a part of the equation:

$$C_T * V_T = C_S * B_{W/S} * (P * V_{W/S} + V_{CITM}) + C_U * B_{W/O} * P * V_{W/O}$$

where C_T Includes the monthly fee for the different seasons.

A set of graphs for both systems based on the number of buildings and fire fees for sprinklered and unsprinklered fees can be found in Appendix K. The volume of water used to develop the graphs was derived from IFC and NFPA 13, 13D, 13R and actual water data taken from the reports discussed.

9.7 Conclusion

As shown in each of the calculated data sheets and graphs generated in the appendices, the fees for sprinklered and unsprinklered buildings are unique in how each of the jurisdictions currently charges for water use. However, each of the jurisdictions can be adapted to use the same basic equation as fees are re-distributed amongst sprinklered and unsprinklered buildings.

$$C_T * V_T = C_S * B_{W/S} * (P * V_{W/S} + V_{CITM}) + C_U * B_{W/O} * P * V_{W/O}$$

In all cases, when a building is provided with a fire protection system, the amount of water used to fight the fire is less than that of buildings without a fire protection system when comparing a single building.

The commodity of water is being sold to the end user. By virtue of this study, the only time water is used, is during a fire or during CITM. Therefore, if both are accounted for by volume, and fire service departments will use the water regardless of whether a building is protected by a sprinkler system or not, then the water used should be distributed to both buildings and not borne by the owner that provides a means for reducing the water used.

Chapter 10
American Water Works Association (AWWA)

American Water Works Association provides guidance on calculating fire protection fees through their document Chapter IV.8 "Rates for Fire Protection Service"[26]. The document provides useful information regarding: issues, history, defining "public" and "private" fire protection and several means for calculating fire protection costs.

Basically there are two methods for calculating fire protection fees:

- Base—extra capacity method
- Commodity—demand cost allocation

It is important to understand that AWWA recognizes that water used for fire protection should be accounted for by the purveyor. However, each of the calculations presented do not provide a means of allocating fees based on the potential water used to fight a fire when a building is provided with a fire protection system.

This report is not intended to replace the means and methods of calculating the total water or costs associated with fire protection water. The means for calculating total costs are still valid and can be used to determine total fees. The equations within this report would be used to supplement or further define the cost allocation when a sprinkler system is in place.

For instance, the example provided by AWWA, Table 10.1 "Customer class fire flow demands and unit cost—Base extra capacity method (test year)" below; represents the public fire service cost allocation for the base-extra capacity method.

The total cost for residential fees can be calculated by taking the Total Fire Protection Cost and multiplying it by the ratio of the number of Residential Equivalent Fire Service Demands by the Total Inside City Equivalent Fire Service Demands:

$$\$516,180.00 * (1,857,600/2,326,200) = \$412,198.00$$

Code Consultants, Inc., *Fire Flow Water Consumption in Sprinklered and Unsprinklered Buildings*, SpringerBriefs in Fire, DOI: 10.1007/978-1-4614-8109-6_10, © Fire Protection Research Foundation 2012

Table 10.1 IV.8-3 Customer class fire flow demands and unit cost—base extra capacity method (test year)

Line no.	Customer class	Maximum needed fire Flow gpm	Duration minutes	Number of customers	Equivalent fire service demands 1,000 gallons
	Inside city: Retail service				
1	Residential	1,000	120	15,480	1,857,600
2	Commercial	2,000	180	1,220	439,200
3	Industrial	3,500	240	35	29,400
4	Total inside city fire protection units			16,735	2,326,200
5	Total fire protection cost				$516,180
6	Total inside city fire protection units				2,326,200
7	Unit cost, $/unit				$0.2219

The information is then plugged into the equation provided in Section XI as Total Cost. The information plugged into the spreadsheet in Appendix L to calculate the fee per customer is as follows:

* Total buildings   15,480
* Probability   0.35 %
* Line Size   1"
* Fee Cap   N/A
* Fees Received Currently $412,199.00

Appendix L provides two examples of the calculation for a residential occupancy. The first example shows the fees for sprinklered and unsprinklered buildings without a capped fee associated with the unsprinklered building. The graphic illustration indicates that as the number of sprinklered buildings increase, the price of the unsprinklered building increases. This stands to reason since the amount of water used in an actual fire in a sprinklered building is 90 % less than an unsprinklered building. Since the total water produced has to be accounted for in the fees identified by AWWA calculations, the fees for the water not used are shifted to the unsprinklered buildings.

The second example provides an imaginary cap identified at $5.00 for an unsprinklered property. When the cap is reached, the fees are then redistributed to the sprinklered buildings.

The model used in this exercise expands the results of the AWWA example providing an avenue of differentiating between sprinklered and unsprinklered buildings. The same calculations can be performed on private mains and added to both public and private fees.

References

1. International Code Council. International fire code. Country Club Hills: International Code Council, Inc.; 2012.
2. National Fire Protection Association. NFPA 1, fire code. Quincy: National Fire Protection Association; 2012.
3. National Fire Protection Association. NFPA 13, standard for the installation of sprinkler systems. Quincy: National Fire Protection Association; 2010.
4. Insurance Services Office. Guide for determination of needed fire flow. Jersey City: ISO Properties, Inc.; 2008.
5. National Fire Protection Association. NFPA 13D, standard for the installation of sprinkler systems in one and two family dwellings and manufactured homes. Quincy: National Fire Protection Association; 2010.
6. National Fire Protection Association. NFPA 13R, standard for the installation of sprinkler systems in residential occupancies up to and including four stories in height. Quincy: National Fire Protection Association; 2010.
7. Water Research Foundation. Residential end uses of water. Denver: AWWA Research Foundation and the American Water Works Association; 1999.
8. Barber NL, Hutson SS, Linsey KS, Lovelace JK, Maupin MA, Kenny JF. "Estimated use of water in the United States in 2005," U.S. Geological Survey Circular 1344. 2009. 52 p.
9. U.S. Census Bureau. American community survey. [Online]. (2005–2009). http://www.census.gov/acs/www/
10. U.S. Census Bureau. Economic census. [Online]. (2007). http://www.census.gov/econ/census07/
11. The Village of Johnstown Ohio. Utilities Department. [Online]. (2011). http://www.villageofjohnstown.org/utilities.html
12. Missouri City of Saint Charles. Public Works Fees. [Online]. (2011). http://www.stcharlescitymo.gov/Departments/PublicWorks/Fees/tabid/456/Default.aspx
13. OUC The Reliable One. Rates. [Online]. (2011). http://www.ouc.com/en/news_and_information_ctr/rates.aspx
14. New York City of Rochester. Rates for water consumption and services. [Online]. (2011). http://www.cityofrochester.gov/article.aspx?id=8589937103
15. Denver Water. 2011 water rates. [Online]. (2011). http://www.water.denver.co.gov/BillingRates/RatesCharges/2011Rates/InsideCity/
16. LA Department of Water & Power. Water Rates. [Online]. (2011). http://www.ladwp.com/ladwp/cms/ladwp001155.jsp
17. International Code Council. International building code. Country Club Hills: International Code Council; 2012.

18. National Fire Protection Association. NFPA 101 life safety code. Quincy: National Fire Protection Association; 2012.
19. National Fire Protection Association. NFPA 25, standard for the inspection, testing, and maintenance of water-based fire protection systems. Quincy: National Fire Protection Association; 2011.
20. U.S. Census Bureau. Census of Housing. [Online]. (2011). http://www.census.gov/hhes/www/housing/census/historic/units.html
21. U.S. Census Bureau. 2009 nonemployer statistics. [Online]. (2009). http://censtats.census.gov/cgi-bin/nonemployer/nonsect.pl
22. National Fire Protection Association. Structure fires by occupancy. Quincy: National Fire Protection Association; 2011.
23. Utiskul Y, Wu N. "Residential Fire Sprinklers—Water Usage and Meter Performance Study," The Fire Protection Research Foundation, Quincy, Final Report 2011
24. CSP, FSF P.E., Greg Jakubowski P.E. Communities with home fire sprinklers: the experience in bucks County, Pennsylvania. Washington Crossing: Fire Planning Associates, Inc.; 2011.
25. Hall Jr. JR. U.S. experience with sprinklers. Quincy: National Fire Protection Association; 2011.
26. American Water Works Association (AWWA). (2010) www.drinktap.org. [Online]. 2010. http://www.drinktap.org/consumerdnn/Home/WaterInformation/Conservation/WaterUseStatistics/tabid/85/Default.aspx